YOUR KNOWLEDGE HAS VALUE

Michael Dienst

Local Search with Progress Spectrum Adaptation

GRIN Verlag

Bibliografische Information der Deutschen Nationalbibliothek:

Die Deutsche Bibliothek verzeichnet diese Publikation in der Deutschen National-bibliografie; detaillierte bibliografische Daten sind im Internet über http://dnb.d-nb.de/ abrufbar.

Imprint:

Copyright © 2012 GRIN Verlag GmbH
Druck und Bindung: Books on Demand GmbH, Norderstedt Germany
ISBN: 978-3-656-26354-8

GRIN - Your knowledge has value

Der GRIN Verlag publiziert seit 1998 wissenschaftliche Arbeiten von Studenten, Hochschullehrern und anderen Akademikern als eBook und gedrucktes Buch. Die Verlagswebsite www.grin.com ist die ideale Plattform zur Veröffentlichung von Hausarbeiten, Abschlussarbeiten, wissenschaftlichen Aufsätzen, Dissertationen und Fachbüchern.

Visit us on the internet:

http://www.grin.com/

http://www.facebook.com/grincom

http://www.twitter.com/grin_com

Dipl.-Ing. Michael Dienst

University of Applied Sciences Berlin, Germany
BIONIC RESEARCH UNIT

Local Search with Progress Spectrum Adaptation.

Search algorithms with intergenerational information utilization are considered efficient optimization strategies. Core mechanism is the adaptation of process parameters. However, the costs of data and declaration of traditional strategies are high. With the transfer of adaptation processes in the spectral range of the object variables, a very elegant and efficient algorithm appears. The paper explores the convergence behavior of processing simple but high-dimensional quality functions.

INTRO. Evolutionary Algorithms (EA) are local search methods. They use mechanisms of biological evolution to solve high-dimensional numerical optimization problems [Her-00] [Her-05] [Kah-91] [Kos-03] [Rec 94] [Schw95]. Among them are Genetic Algorithms (GA) and Evolution Strategies (ES).

The code of an Evolutionary Strategy as very compact, the process flow is simple: First, copies of an startup artificial system will be made. Random modifications lead to a multitude of variants of the ELTER- system (variation). MUTANTs and ELTER form a common ensemble of selection. In each generation, all variations of the current ELTER assessed using an objective function and determines the quality of all systems (evaluation). From the crowd, a new weighted system, the current ELTER for the next generation is chosen (selection). The variation of the ELTER system is continuing the campaign. In this manner, the quality of the ensemble rise from generation to generation.
Evolution Strategies are the subject of this paper.

At the local investigation of a complex quality landscape, the number of relevant simulation function calls is important. In order to optimize the industrial practice to be interesting at all, the aim of optimization algorithms development is to reduce the variation and the condition of the ensemble. A proven method to accelerate the convergence of local search is the inclusion of past object variables [Ost-97] [Han-98] [Rec-94] [Lev-95]. However, the expenditure of declaration and data becomes more complex. "Richtungslernen" (Schwefel) and "Präteritum-Strategie" (Rechenberg) the declaration expenditure rises linear, "Covariance Matrix Adaptation, CMA"(Hansen) square and "Erzeugendensystem-Adaption ESA" (Ostermeier) the declaration expenditure rises cubically with the dimension of the optimization problem.

FSA. A local search algorithm using the across generation information for progress spectrum adaptation is described by the author in [Die-12]. The core mechanism of the "Progress Spectrum Adaptation" (*german: Fortschritt Spektren Adaptation, FSA*) mentioned method is the transformation process in their spectral data, their processing, analysis and compression, and inverse - transformation to the functional area of the optimization process. The further processing of the information of progress of the object variables in the spectral range results in a generalization of the random number distribution of the variant form in the functional area and leads to a trajector of the object variables in the optimization.

The Progress in time (n) of an optimization campaign is the difference between the object variable vector of the ELTER **V**e (n, m) of the previous generation (n-1) and the object variable vector **V**b (n-1, m) of the recent (n) best descendant.

The spectral gradient Δ**S**(n) is the difference of the Fourier Transformed of these two vectors to see in the form(1). The spectral gradient Δ**S**(n, m) in the generation (n) and a current vectorial random spectrum **R(n,m)** = FT{ (Z(m)) } has the dimension (m) of the object variable vector of the optimization campaign so that the object variable vector $V(n+1,m)$ of the following generation in the form (2) can be represented. $\delta(n)$ is the hereditary global mutation-step-size parameter which is the orthogonal inverse transformation of iFT spectral range (transform domain) in the local region (object space), as described in [Die-12].

$$\Delta \underline{S}(n,m) = \Delta [FT\{ (Vb(n-1,m)) \} , FT\{ (Ve(n,m)) \}] \qquad (1)$$

$$\underline{V}(n+1,m) = \underline{V}(n,m) + \delta(n) \; iFT\{ (\Delta \underline{S}(n,m) + \underline{R}(n,m)) \} \qquad (2)$$

The presented strategy (FSA) gets information from the analysis of vector optimization progress and adaptation to spectral level by this vector with a random number distribution entangled. An orthogonal inverse transformation back to the image area of the object variables generates a mutational variation distribution in each generation of the optimization campaign.

Model features and simulation experiments. Model functions in (Table 1). The number of function calls is limited in each generation. Made against over-the convergence of evolution strategies (gES) with global mutational step-size control and progress spectrum adapting algorithms (FSA).

University of Applied Sciences Berlin, Germany
BIONIC RESEARCH UNIT

Table 1. Model functions for optimization
experiments. Linie, Ebene, Kubus,
Sphäre. Q=Σ F(x) ➔ Min.

```
function q=Line(x);
q=0.0; dim=length(x);
for i=1:dim q=q+abs((dim/i)-x(i)^1);    end;
endfunction;

function q=pane(x);
q=0.0; dim=length(x);
for i=1:dim q=q+abs((dim/i)-x(i)^2);    end;
endfunction;

function q=cube(x);
q=0.0; dim=length(x);
for i=1:dim q=q+abs((dim/i)-x(i)^3);    end;
endfunction;

function q=spac(x);
q=0.0; dim=length(x);
for i=1:dim q=q+abs((dim/i)-x(i)^4);    end;
endfunction;
```

A comparison of the model calculation of a cubic function with a 100-dimensional object variable vector (CUBE (100)) of the described algorithm with Progress Spectrum Adaption (FSA) with a classical evolution strategy with global mutation step size control (gES) now shows the following results: The quality function q of the algorithm with local FSA is clearly beneficial in the early stage of the optimization campaign. The processing using the across generation information leads to a higher orientation, which is especially true when the structure optimization problem is still bad (in the early phase of the optimization, Fig.1). With progress optimization the classical evolutionary strategy converges better.

Temporal development of object variables. A sensitivity analysis with the observation of the optimization process is the difference between the object variable vectors two successive generations, the progress of the local algorithm. A monitoring process utilizes the Euclidean distance of the vectors **Vb**, and **Ve**.

Euclidean distance

$$\text{dist,E } (\underline{Vb}, \underline{Ve}) \;=\; [\, \Sigma \,[\, \underline{V}b(n\text{-}1,m) - \underline{V}e(n,m) \,]^2 \,]^{(1/2)} \qquad (3)$$

The Euclidean distance, of the vectors Vb and Ve is structurally related to the variance of a difference vector. If enough information is known about the development of two vectors is the Euclidean distance an efficient criterion for the quality of local progress.

The calculation (Fig. 2) shows the typical course of an optimization campaign on a bi-quadratic function with 100-dimensionelem object variable vector for an evolution strategy with processing using the across generation information. The Euclidean distance of the object variable development has in the "early stage" of the campaign, their maximum values, however, has for the SPAC function is not the initial acceleration, which is observed in the linear, quadratic and cubic test functions (Figure 3).

The effect of the adaptation of the progress spectrum of the object variable development of the quality of the diagrams in the course of an optimization campaign and the Euclidean distances of progress. The adaptation effect is particularly strong in the early stage of the optimization campaign. In practice, therefore, to optimize an algorithm would be desirable that the strategy paradigm of the early stage, the orientation distribution of the mutation spectrum in progress, gives up in favor of a stochastic variant formation in the convergence of the optimization. In light of this, we are continuing our research on local search strategies with across generation information processing.

Local Search with Progress Spectrum Adaptation

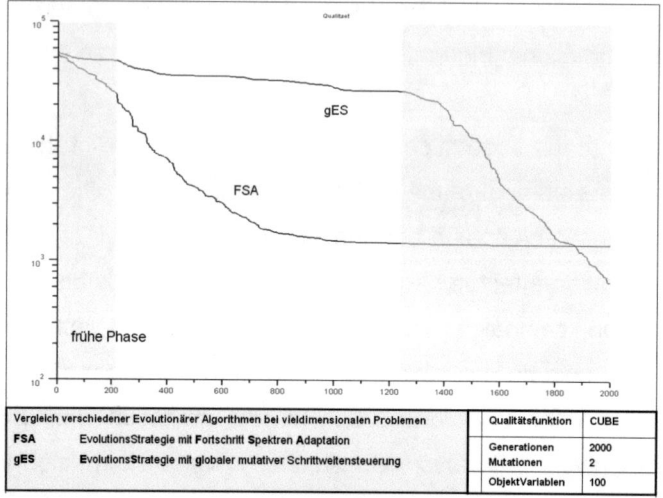

Vergleich verschiedener Evolutionärer Algorithmen bei vieldimensionalen Problemen		Qualitätsfunktion	CUBE
FSA	EvolutionsStrategie mit Fortschritt Spektren Adaptation	Generationen	2000
gES	EvolutionsStrategie mit globaler mutativer Schrittweitensteuerung	Mutationen	2
		ObjektVariablen	100

Figure1: CUBE(100) (Q=Σ F(x) ➜ Min.)

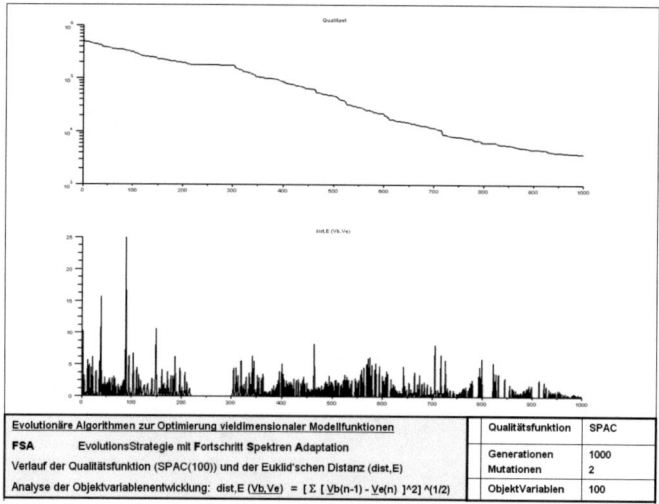

Evolutionäre Algorithmen zur Optimierung vieldimensionaler Modellfunktionen		Qualitätsfunktion	SPAC
FSA EvolutionsStrategie mit Fortschritt Spektren Adaptation		Generationen	1000
Verlauf der Qualitätsfunktion (SPAC(100)) und der Euklid'schen Distanz (dist,E)		Mutationen	2
Analyse der Objektvariablenentwicklung: dist,E ($\underline{Vb},\underline{Ve}$) = [Σ [$\underline{Vb}$(n-1) - $\underline{Ve}$(n)]^2] ^(1/2)		ObjektVariablen	100

Figure 2: Function SPAC(100) (Q=Σ F(x) ➜ Min.)

University of Applied Sciences Berlin, Germany
BIONIC RESEARCH UNIT

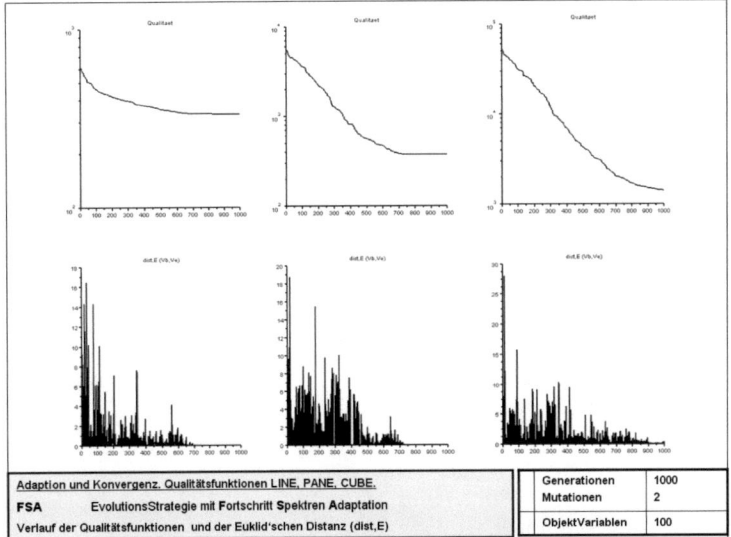

Adaption und Konvergenz. Qualitätsfunktionen LINE, PANE, CUBE.		Generationen	1000
FSA EvolutionsStrategie mit Fortschritt Spektren Adaptation		Mutationen	2
Verlauf der Qualitätsfunktionen und der Euklid'schen Distanz (dist,E)		ObjektVariablen	100

Figure3: funktion Linie, Ebene, Kubus, Sphäre $(Q=\Sigma\ F(x))$ ➔ **Min.**)

Bibliographie

[Die-12]   Dienst, M., (2012) Optimierung mit generationsübergreifender Informationsausnutzung. In Forschungsbericht 2011 der BHT Berlin, S. 179-183. Publikationen der Beuth Hochschule für Technik Berlin. ISBN 978-3-856-73650-5.

[Han-98]   Hansen, N. (1998) Verallgemeinerte individuelle Schrittweitenregelung in der Evolutionsstrategie. Dissertation, Technische Universität Berlin 1998.

[Her-00]   Herdy, Michael, (2000) Beiträge zur Theorie und Anwendung der Evolutionsstrategie. Mensch und Buch Verlag, Berlin.

[Her-05]   Herdy, Michael, (2005) Anwendung der Evolutionsstrategie in der Industrie. In Evolution zwischen Chaos und Ordnung. S. 123 – 138. Freie Akademie Verlag, Bernau.

[Kos-03]   Kost, Bernd, (2003) Optimierung mit Evolutionsstrategien. Harri Deutsch Verlag, Frankfurt a. M.

[Mef-04]   Meffert, B., Hochmut, O. (2004) Werkzeuge der Signalverarbeitung. Pearson-Studium, München.

[Ost-97]   Ostermeier, A. (1997) Schrittweitenadaptation in der Evolutionsstrategie mit einem entstochastisierten Ansatz. Diss. Technische Universität Berlin 1997.

[Rec-94]   Rechenberg, Ingo, (1994) Evolutionsstrategie. Frommann Holzboog Verlag Stuttgart- Bad Cannstatt.

[Sche-85] Scheel, Armin (1985) Beitrag zur Theorie der Evolutionsstrategie. Dissertation, TU Berlin.

[Schw-95] Schwefel, H.–P. (1995) Evolution and Optimum Seeking. John Wiley & Sons. New York.